Stars So Bright:
Book of Constellations (Kiddie edition)

Speedy Publishing

Speedy Publishing LLC
40 E. Main St. #1156
Newark, DE 19711
www.speedypublishing.com

Copyright 2015

All Rights reserved. No part of this book may be reproduced or used in any way or form or by any means whether electronic or mechanical, this means that you cannot record or photocopy any material ideas or tips that are provided in this book

Constellations are basically groups of stars that have imaginatively been linked together to depict mythological characters, animals and objects from mankind's past.

In 1922, the International Astronomical Union (IAU) officially recognized 88 constellations.

Aries is one of the constellations of the zodiac. It is located in the northern celestial hemisphere between Pisces to the west and Taurus to the east.

Cancer is one of the twelve constellations of the zodiac. Its name is Latin for crab and it is commonly represented as one.

Taurus is one of the constellations of the zodiac, which means it is crossed by the plane of the ecliptic.

Pisces is a constellation of the zodiac. Its name is the Latin plural for fish.

LIBRA

Libra is a constellation of the zodiac. Its name is Latin for weighing scales, and its symbol is Libra.

Capricornus is one of the constellations of the zodiac. Its name is Latin for "horned goat" or "goat horn", and it is commonly represented in the form of a sea-goat: a mythical creature that is half goat, half fish.

CAPRICORN

Aquarius is a constellation of the zodiac, situated between Capricornus and Pisces. Its name is Latin for "water-carrier" or "cup-carrier", and its symbol is Aquarius, a representation of water.

Leo is one of the constellations of the zodiac, lying between Cancer to the west and Virgo to the east. Its name is Latin for lion.

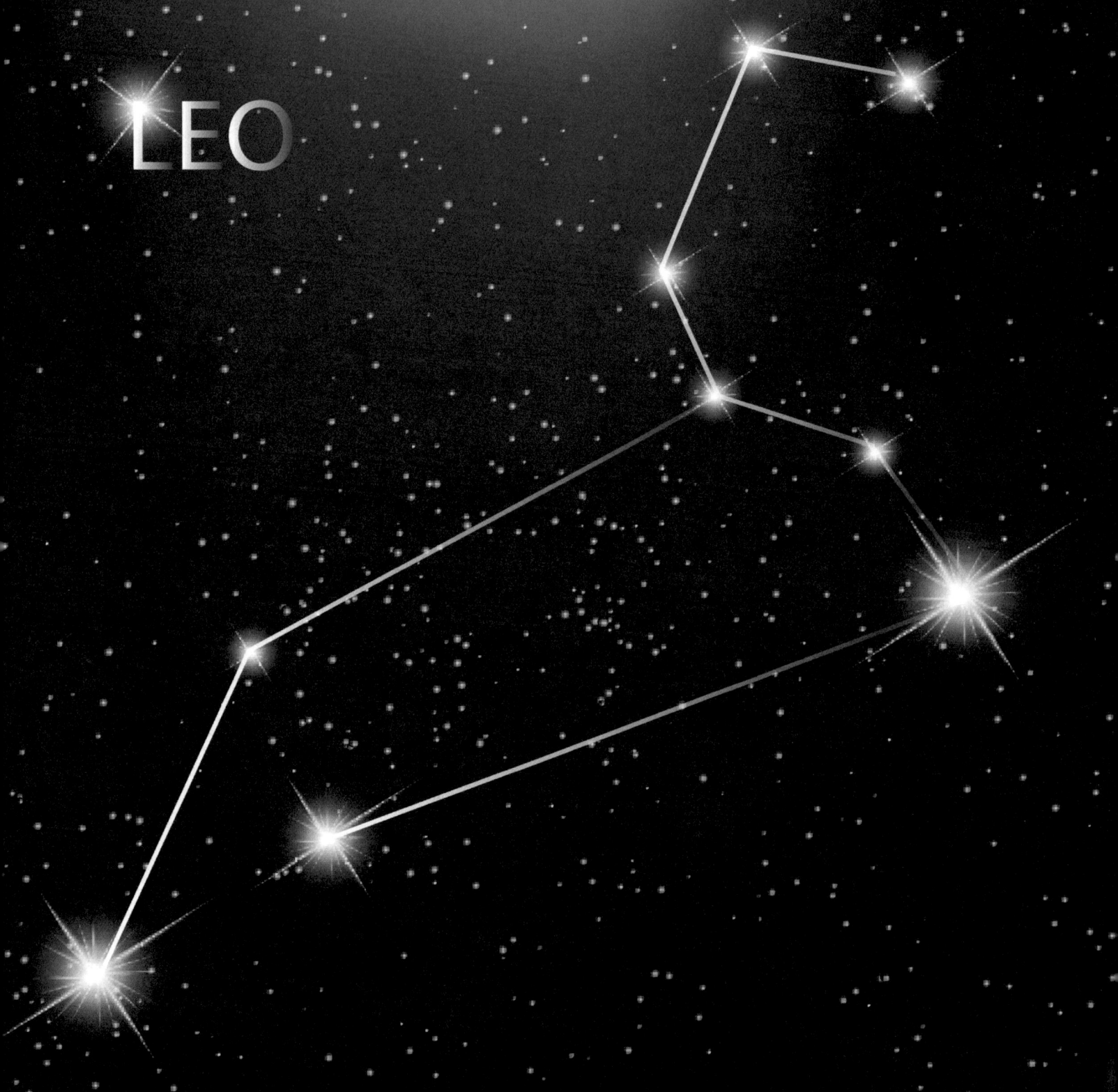

Scorpius is one of the constellations of the zodiac. Its name is Latin for scorpion, and its symbol is Scorpio.

Sagittarius is one of the constellations of the zodiac. Sagittarius is commonly represented as a centaur drawing a bow.

GEMINI

Gemini is one of the constellations of the zodiac. It is associated with the twins Castor and Pollux in Greek mythology.

Virgo is one of the constellations of the zodiac. Its name is Latin for virgin.

Throughout the year different constellations become visible to us in the night sky as the Earth completes its annual orbit around the Sun.

Made in the USA
Middletown, DE
18 February 2017